BATS BENEATH THE BRIDGE

JANET NOLAN

ILLUSTRATED BY EMILY COX

Albert Whitman & Company
Chicago, Illinois

To Bill, Tom, and Megan, who saw the bats fly.—JN

For all the young readers who let their imaginations fly free.—EC

Library of Congress Cataloging-in-Publication data is on file with the publisher.

Illustrations by Emily Cox
First published in the United States of America in 2024 by Albert Whitman & Company
ISBN 978-0-8075-0562-5 (hardcover)
ISBN 978-0-8075-0563-2 (ebook)

Printed in China

10 9 8 7 6 5 4 3 2 1 WKT 28 27 26 25 24

Design by Shane Tolentino and Erin McMahon

For more information about Albert Whitman & Company,
visit our website at www.albertwhitman.com

FOREWORD

DR. MERLIN TUTTLE

This is a true story of what the people of Austin, Texas, learned about bats. Bats may look scary, but they leave people alone unless we disturb them. Bats help us by eating insects, so we don't need to use as many pesticides as we would without them. Pesticides aren't healthy for people.

There are more than one thousand four hundred kinds of bats, each with its own characteristics. Bumblebee bats weigh less than one penny. Some flying fox bats have wingspans of nearly six feet. Painted bats are brightly colored. Free-tailed bats adopt orphaned bats and raise them as their own.

In tropical areas, bats are often the best pollinators of flowers. They carry pollen farther than any other pollinator. Without bats, we might not have bananas; banana flowers open at night to attract the bats that pollinate them. Cashew nuts and figs also grow on trees that rely on bats to carry their seeds.

Bats are very smart. Scientists can train them to fly to a specific place or to come back, similar to dogs. Bats remember what they're taught and can even teach other bats. They have special bat friends that they travel with, sharing information.

I hope you'll want to learn more about these amazing but often misunderstood animals.

In early spring, a bat arrived in Austin, Texas.

"The bats are back!"

Its long migration from its winter home in Mexico was over.

The tired bat found a cozy crevice under the bridge and crawled inside. It tucked in its wings, hooked its claws into the concrete, and fell asleep, hanging upside down.

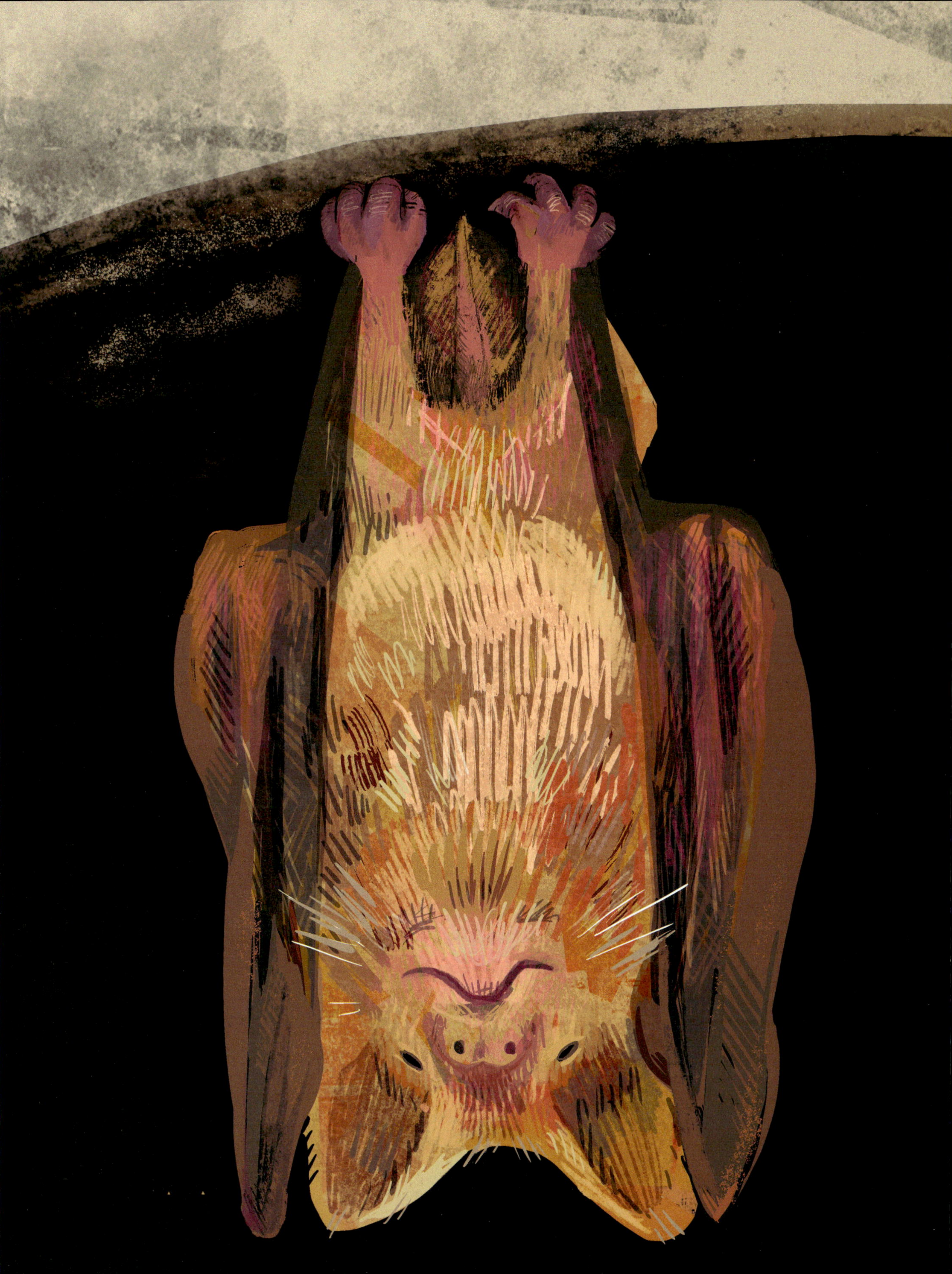

The bat slept until the sun began to set.

Then it flew high into the sky.

But it was not alone.

One bat…two bats…fifty bats…hundreds of bats flew out from a bridge that had been closed for repairs.

“The bats are back!”

Everybody had thought the rebuilt bridge was just a bridge.

But it wasn't.

The engineers who designed the repairs didn't know.

The bridge builders who repaired the bridge didn't know.

Nobody knew the long, thin crevices beneath the bridge were the perfect size and shape for Mexican free-tailed bats.

Bats started moving in, and not just a few.

Ten thousand...fifty thousand...one hundred thousand bats began roosting beneath the Congress Avenue Bridge.

Nocturnal, the bats slept during the day. At dusk they woke, and then they flew out from the bridge and over the city on their way to the farmers' fields to hunt their prey.

People in the city were worried about bats in their bridge. They wrote letters to the newspaper.

"Those bats are scary."

"We don't want them to stay."

That's when a bat biologist came to town. Dr. Merlin Tuttle visited schools to get the word out: "Bats are good for the environment."

Bats eat insects, lots of insects. Moths and beetles that harm crops. And mosquitoes that bite!

WHAT BATS EAT

More bats moved in. Two...three... four...five hundred thousand bats roosted inside the bridge.

Some people in the city were still concerned about bats in their bridge. They spoke up at meetings.

"Those bats are frightening."

"We don't want them to stay."

But now, not everyone felt that way.

"We want the bats to stay."

In the evening, as the sun began to set, bat watchers came to the bridge to see the bats fly.

They watched and waited.

And then...

"Look!" they shouted.

"There!" they yelled.

Bats and more bats, the only mammals that can fly, emerged from the bridge. Thousands of flapping bat wings made a soft whooshing sound.

The fast-flying bats stuck close together for protection against predators—hawks and owls. Then they fanned out, flying in smaller and smaller groups until they were flying alone. The bats ate while flying, and they could fly all night in search of food.

Dr. Tuttle met with community groups and city leaders to get the word out. “Bats are good for the environment. Each insect a bat eats is one less pest to destroy the crops that grow in the farmers’ fields.”

More bats moved in. Six…seven…eight…nine hundred thousand bats roosted inside the bridge.

A few people in the city still complained about the bats under their bridge. They wrote their names on petitions.

"Those bats are disgusting."

"We don't want them to stay."

But now, more people were starting to understand the environmental benefits of bats.

Night after night, as the sun dipped lower in the western sky, crowds of bat watchers came to the bridge to see the bats fly.

They watched and waited.

"When will they fly?"

And then...

"Look!" they shouted.

"There!" they yelled.

The bats formed a long, black ribbon of flapping movement that stretched out across the dimming sky.

Bats do not use their eyes to hunt in the dark. They use vibrations in a technique called echolocation. A bat emits a high-pitched peep, too high for humans to hear. When the sound hits a moth or beetle it bounces back into the bat's ear as an echo. The sound tells the bat where to catch its prey.

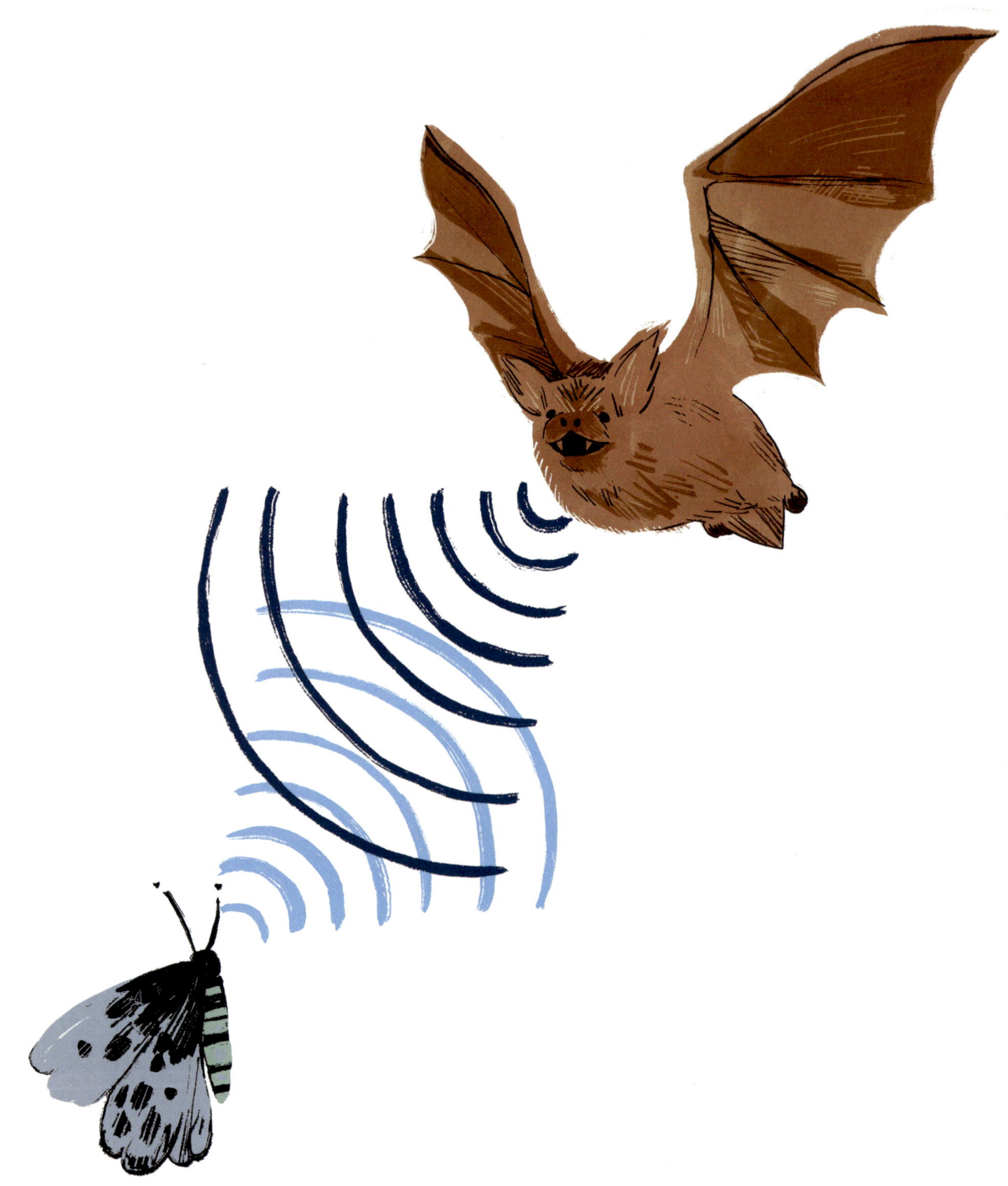

As daylight broke, their bellies full, the bats returned to their roost beneath the bridge. The tired bats tucked in their wings, gripped with their feet, and fell asleep, hanging upside down.

Dr. Tuttle did television and radio interviews to get the word out: "Bats are good for the environment."

In a single night, the bats that lived beneath the bridge could eat up to twenty thousand pounds—ten tons—of insects. Fewer pests meant farmers could use fewer pesticides.

More bats moved in. One million one...two...three...four hundred thousand bats now roosted inside the bridge.

But now the bats were welcome.

The people in the city understood bats are good for people, good for farmers, and good for the environment.

They wanted them to stay!

"Hooray! We love bats!"

In the light of the setting sun, bat watchers from around the city, the country, and beyond now come to the Congress Avenue Bridge, home to the largest urban bat colony in the world. They come at nightfall to see the bats fly.

People paddle onto the water in kayaks and canoes. They buy tickets for a ride on a bat tour boat. Families bring picnics and lay blankets on the grass. They fill the bridge from one end to the other.

They watch and wait.

And then...

"Look! There they are."

In the fading light, one and a half million bats fill the darkening sky.

"Fly, bats, fly!"

HISTORY OF THE CONGRESS AVENUE BRIDGE BATS

An unintended consequence of a bridge repair in the 1980s was that Austin, Texas, became known as the "Bat Capital of America," home to the largest urban bat colony in the world.

The Congress Avenue Bridge is primarily a maternity colony. Most of the bats that return to the bridge in early spring are pregnant females. Each gives birth to one bat, called a pup, usually during the first week of June.

When the mother bats fly at night, they leave the pups behind in the concrete crevices. The baby bats cluster together, approximately four hundred bats per square foot. The mothers return to the bridge before daybreak and find their own pup to nurse.

The bats migrate south in the winter to caves in north and central Mexico. When they return in the spring, so do the people who come to see the bats fly. More than one hundred thousand tourists now come to Austin every year to see the bats fly, adding millions of dollars to the city's economy.

Dr. Merlin Tuttle is a bat biologist and founder of Bat Conservation International, which he led for over thirty years, leaving there in 2009 to start Merlin Tuttle's Bat Conservation in 2014.

BAT FACTS

The Austin, Texas, bridge bats are commonly referred to as Mexican free-tailed bats. Their official name is Brazilian free-tailed bats. They are called free-tailed because their tails extend freely beyond their tail membranes, which are thin, soft layers of skin.

Babies are born without fur and with their eyes open. The mothers stay with their babies for the first hour or so of life, each one learning its baby's smell and the sound of its cry.

Young bats learn to fly at four to five weeks. They find their own food at about six weeks and are on their own by week eight.

Brazilian free-tailed bats weigh about 0.43 ounces and their wingspan is eleven to twelve inches. Their fur is grey to reddish brown, and their lifespan is believed to be about ten years.

Bats are the only mammals that can fly. They can travel up to 60 miles per hour and at heights above 10,000 feet.

Brazilian free-tailed bats mostly eat agricultural pests. They don't use their eyes to hunt in the dark—they use sound waves, vibrations, in a process called echolocation. The bats make high-pitched beeps, too high for humans to hear. When the sound hits its target, an insect, or anything else, it bounces back into the bat's ear as an echo. The sound lets the bat know where to find its prey.

If you find a bat on the ground or if one gets inside a house, call an adult to remove it.

Merlin Tuttle's Bat Conservation is helping to save millions of bats, protecting public health, economies, and worldwide ecosystems by teaching people to live harmoniously with bats. Learn more at www.merlintuttle.org

Bat Conservation International is devoted to conservation, education, and research initiatives involving bats and the ecosystems they serve. Learn more at www.batcon.org

GLOSSARY

migration: When animals, fish, birds, or insects travel from one location to another for food or to breed.

nocturnal: Active at night.

pesticides: Chemicals used to kill or control unwanted pests that can harm gardens, crops, and animals, including people.

pollination: A process that allows plants to make seeds and reproduce. Animals (such as bats, insects, and birds) and wind carry pollen between flowering plants.

predator: An animal that hunts another animal for food.